AN ALMOST PRACTICAL STEP TOWARD SUSTAINABILITY

ROBERT SOLOW

Institute Professor of Economics,
Massachusetts Institute of Technology;
Nobel Laureate in Economic Science

RESOURCES
FOR THE FUTURE

An Invited Lecture on the
Occasion of the Fortieth Anniversary
of Resources for the Future

Presented in the Resources and Conservation Center
New York • October 8, 1992

Published 1992 by Resources for the Future
2 Park Square, Milton Park, Abingdon, Oxon, OX14 4RN
711 Third Avenue, New York, NY 10017

RESOURCES FOR THE FUTURE

Resources for the Future (RFF) is an independent nonprofit organization engaged in research and public education on natural resource and environmental issues. Its mission is to create and disseminate knowledge that helps people make better decisions about the conservation and use of their natural resources and the environment. RFF neither lobbies nor takes positions on current policy issues.

Because the work of RFF focuses on how people make use of scarce resources, its primary research discipline is economics. However, its staff also includes social scientists from other fields, ecologists, environmental health scientists, meteorologists, and engineers. Staff members pursue a wide variety of interests, including forest economics, recycling, multiple use of public lands, the costs and benefits of pollution control, endangered species, energy and national security, hazardous waste policy, climate resources, and quantitative risk assessment.

Acting on the conviction that good research and policy analysis must be put into service to be truly useful, RFF communicates its findings to government and industry officials, public interest advocacy groups, nonprofit organizations, academic researchers, and the press. It produces a range of publications and sponsors conferences, seminars, workshops, and briefings. Staff members write articles for journals, magazines, and newspapers, provide expert testimony, and serve on public and private advisory committees. The views they express are in all cases their own, and do not represent positions held by RFF, its officers, or trustees.

Established in 1952, RFF derives its operating budget in approximately equal amounts from three sources: investment income from a reserve fund, government grants, and contributions from corporations, foundations, and individuals. (Corporate support cannot be earmarked for specific research projects.) Some 45 percent of RFF's total funding is unrestricted, which provides crucial support for its foundational research and outreach and educational operations. RFF is a publicly funded organization under Section 501(c)(3) of the Internal Revenue Code, and all contributions to its work are tax deductible.

CONTENTS

INTRODUCTORY REMARKS

PAUL R. PORTNEY
Vice President
Resources for the Future

Good afternoon and welcome to the Resources and Conservation Center. I'm Paul Portney, vice president of Resources for the Future, and I'm delighted to welcome you to this afternoon's festivities.

As most of you know, RFF is, first and foremost, a research organization, and we take great pride in the quality and relevance of the research done here for the past forty years. What some of you may not know is that we consider ourselves an educational organization, too, albeit one without students, and that we treat this educational mission as seriously as we do our research program. For this reason, therefore, we mail out four times a year—free of charge—our publication *Resources* to more than 25,000 policymakers, academics, members of the business and environmental communities, the press, and interested citizens. What all these readers have in common, we believe, is a desire for carefully reasoned, clearly written, and impartial analysis of natural resource and environmental policy issues. We try awfully hard to satisfy that desire.

We try to fulfill our educational mission in other ways, as well, ranging from book publication to grant making and from conference sponsorship to the preparation of congressional testimony. We also run a Wednesday seminar program which people attend to hear about new technical developments in, say, the theory of benefit-cost analysis, as well as to become informed—and sometimes provoked—about current policy battles raging in the natural resources and environmental field.

Which brings me to this afternoon's get together. We hope that one of the consequences of these educational programs is to fix in people's minds the thought that RFF is *the* place to turn to for all those economists, scientists, lawyers, policymakers, businessmen and women, environmental advocates, reporters, and others interested in scholarly research and creative policy analy-

1

sis on environmental and natural resource problems. That's one of the reasons why I'm so pleased to have the crème de la crème of the environmental policy community here this afternoon. When planning our fortieth anniversary celebration, therefore, it was natural for us to try to include an event for this audience that would be both enlightening and entertaining, both analytic and applied.

I can report in all honesty that the easiest part of planning the entire anniversary celebration was identifying whom to ask to give such a talk. My clear recollection is that the immediate reaction of all my colleagues was, "Well, obviously we should ask Robert Solow and worry about where to go next only if he says no." Fortunately, we never had that worry.

If your experience is anything like mine, you know the most long-winded introductions inevitably follow the words, "Our next speaker needs no introduction." Well, our next speaker needs no introduction, but I'll try to be brief anyway.

Robert Solow has for the past twenty years been an Institute Professor of Economics at the Massachusetts Institute of Technology. Educated in the New York City public schools, he earned his B.A., M.A., and Ph.D. at Harvard. Along the way he has also been awarded nineteen honorary degrees from universities in six different countries.

In 1950 Professor Solow began his career as an assistant professor in MIT's Statistics Department. In 1958 he became a professor of economics there. He has won the Wells Prize for the best dissertation in economics at Harvard University, the John Bates Clark Medal of the American Economic Association for outstanding contributions by an economist under the age of forty, and—in 1987—was awarded the Alfred Nobel Memorial Prize for Economic Science. He is a member of the National Academy of Sciences, a past vice president of the American Association for the Advancement of Science, and a past president of both the Econometric Society and the American Economic Association. (Can you understand the growing feeling of inadequacy I felt as I read his c.v.?)

Despite this quite remarkable career, if you ask those who know Professor Solow well what they admire most about him, you will likely get an answer that no vita, however long, could include. Rather, you will be told, what really elevates him above the crowd has been the dedication—throughout his career—-with which he has tried simultaneously to expand the technical frontiers in economics, improve the way it is taught, and focus improved analytical methods not on meaningless arcana but on the real-world problems that affect the lives of us all—particularly those less fortunate than ourselves. To put it directly, his career is the embodiment of responsibility to one's profession, one's country, and one's fellow man. This has led him in his research to ponder, among other important subjects, the wise management of society's endowment of natural resources, the subject of his talk this afternoon. I hope it's obvious why we at RFF are so honored to have him here this afternoon. Ladies and gentlemen, Professor Robert Solow . . .

RESOURCES

AN ALMOST PRACTICAL STEP TOWARD SUSTAINABILITY

ROBERT SOLOW

You may be relieved to know that this talk will not be a harangue about the intrinsic incompatibility of economic growth and concern for the natural environment. Nor will it be a plea for the strict conservation of nonrenewable resources, even if that were to mean dramatic reductions in production and consumption. On the other hand, neither will you hear mindless wish fulfillment about how ingenuity and enterprise can be counted on to save us from the consequences of consuming too much and preserving too little, as they have always done in the past.

Actually, the argument I want to make seems to be particularly appropriate on the occasion of the fortieth anniversary of Resources for the Future; it is precisely about resources for the future. And it is even more appropriate for a research organization: I hope to show how some fairly interesting pure economic theory can offer a hint—though only a hint—about a possible improvement in the way we talk about and think about our economy in relation to its endowment of natural resources. The theoretical insight that I will present suggests a potentially important line of empirical research and a possible guideline for long-term economic policy. Then I will make a naive leap and suggest that, if we talked about the economy in a more sensible and accurate way, we might actually be better able to conduct a rational policy in practice with respect to natural and environmental resources. That is probably foolishness, but I hope you will find it a disarming sort of foolishness.

PREVIEWING THE ARGUMENTS

It will be useful if I tell you in advance where the argument is leading. It is a commonplace thought that the national income and product accounts, as currently laid out, give a misleading pic-

5

ture of the value of a nation's economic activity to the people concerned. The conventional totals, gross domestic product (GDP) or gross national product (GNP) or national income, are not so bad for studying fluctuations in employment or analyzing the demand for goods and services. When it comes to measuring the economy's contribution to the well-being of the country's inhabitants, however, the conventional measures are incomplete. The most obvious omission is the depreciation of fixed capital assets. If two economies produce the same real GDP but one of them does so wastefully by wearing out half of its stock of plant and equipment while the other does so thriftily and holds depreciation to 10 percent of its stock of capital, it is pretty obvious which one is doing a better job for its citizens. Of course the national income accounts have always recognized this point, and they construct net aggregates, like net national product (NNP), to give an appropriate answer. Depreciation of fixed capital may be badly measured, and the error affects net product, but the effort is made.

> There is a "right" way to charge the economy for the consumption of its resource endowment and for the degradation or improvement of environmental assets, and those measurements play a central role in the only logically sound approach to the issue of sustainability that I know.

The same principle should hold for stocks of nonrenewable resources and for environmental assets like clean air and water. Suppose two economies produce the same real net national product, with due allowance for depreciation of fixed capital, but one of them is wasteful of natural resources and casually allows its environment to deteriorate, while the other conserves resources and preserves the natural environment. In such a case we have no trouble seeing that the first is providing less amply for its citizens than the second. So far, however, the proper adjustments needed to measure the stocks and flows of our natural resources and environmental assets are not being made in the published national accounts. (The United Nations has been working in this direction for some years, so the situation may change, although only with respect to environmental accounting.) The nature of this problem has been understood for some time, and individual scholars, beginning with William D. Nordhaus and James Tobin in 1972, have made occasional passes at estimating the required corrections.

That is hardly news. The additional insight that I want to explain is that there is a "right" way to make that correction—not perhaps the easiest or most direct way, but the way that properly charges the economy for the consumption of its resource endowment. The same principle can be extended to define the right adjustment that must be made to allow for the degradation or improvement of environmental assets in the course of a year's economic activity. The properly adjusted net national product would give a more meaningful indicator of the annual contribution to economic well-being.

The corrections are more easily defined than performed. The necessary calculations would undoubtedly be more error-prone than those the U.S. Department of Commerce already does with respect to the depreciation of fixed capital. Nevertheless, I would suggest that talk without measurement is cheap. If we— the country, the government, the research community—are serious about doing the right thing for the resource endowment and the environment, then the proper measurement of stocks and flows ought to be high on the list of steps toward intelligent and foresighted decisions.

The second and last step in my argument is more abstract. It turns out that the measurements I have just been discussing play a central role in the only logically sound approach to the issue of sustainability that I know. If "sustainability" is anything more than a slogan or expression of emotion, it must amount to an injunction to preserve productive capacity for the indefinite future. That is compatible with the use of nonrenewable resources only if society as a whole replaces used-up resources with something else. As you will see when I return to this point for a full exposition, the very same calculation that is required to construct an adjusted net national product for current evaluation of economic benefit is also essential for the construction of a strategy aimed at sustainability. This conclusion confirms the importance of a serious effort to dig out the relevant facts.

That is a brief preview of what I intend to say, but before going on to say it, I would like to mention the names of the economists who have contributed most to this line of thought. They include Professors John Hartwick of Queen's University in Canada, Partha Dasgupta of the University of Cambridge, England, and Karl-Göran Mäler of the Stockholm School of Economics; my sometime colleague Martin L. Weitzman, now of Harvard University; and, more on the practical side, Robert Repetto of the World Resources Institute. I have already mentioned the early work of Nordhaus and Tobin; Nordhaus has continued to contribute common sense, realism, and rigorous economic analysis. Finally, I should confess that I have contributed to this literature myself. My idea of heaven is an occasion when a piece of pretty economic theory turns out to suggest a program of empirical research and to have implications for the formulation of public policy.

FINDING THE TRUE NET PRODUCT OF OUR ECONOMY

Now I go back to the beginning and make my case in more detail. Suppose we adopt a simplified picture of an economy living in some kind of long run. What I mean by that awkward phrase is that we are going to ignore all those business-cycle problems connected with unemployment and excess capacity or overheating and inflation. From quarter to quarter and year to year this economy fully exploits the resources of labor, plant, and equipment that are available to it.

To take the easiest case—that of natural resources—first, imagine that this economy starts with a fixed stock of nonrenewable resources that are essential for further production. This is an oversimplification, of course. Even apart from the possibility of exploration and discovery, the stock of nonrenewable resources is not a pre-existing lump of given size, but a vast quantity of raw materials of varying grade, location, and ease of extraction. Those complications are not of the essence, so I ignore them.

It is of the essence that production cannot take place without some use of natural resources. But I shall also assume that it is always possible to substitute greater inputs of labor, reproducible capital, and renewable resources for smaller direct inputs of the fixed resource. Substitution can take place on reasonable terms, although we can agree that it gets more and more costly as the process of substitution goes on. Without this minimal degree of optimism, the conclusion might be that this economy is like a watch that can be wound only once: it has only a finite number of ticks, after which it stops. In that case there is no point in talking about sustainability, because it is ruled out by assumption; the only choice is between a short happy life and a longer unhappy one.

Life for this economy consists of using all of its labor and capital and depleting some of its remaining stock of resources in the production of a year's output (GDP approximately). Part of each year's output is consumed, and that gives pleasure to current consumers; the rest is invested in reproducible capital to be used for production in the future. There are various assumptions one could make about the evolution of the population and employment. I will assume them to have stabilized, since I want to talk about the very long run anyway. Next year is a lot like this year, except that there will be more plant and equipment, if net investment was positive this year, and there will be less of the stock of resources left.

Each year there are two new decisions: how much to save and invest, and how much of the remaining stock of nonrenewable resources to use up. There is a sense in which we can say that this year's consumers have made a trade with posterity. They have used up some of the stock of irreplaceable natural resources; in exchange they have saved and invested, so that posterity will inherit a larger stock of reproducible capital.

This intergenerational trade-off can be managed well or badly, equitably or inequitably. I want to suppose that it is done well and equitably. That means two things. First, nothing is simply wasted; production is carried on efficiently. Second, although the notion of intergenerational equity is much more complicated and I cannot hope to explain it fully here, the idea is that each generation is allowed to favor itself over the future, but not too much. Each generation can, in turn, discount the welfare of all future generations, and each successive generation applies the same discount rate to the welfare of its successors.

9

To make conservation an interesting proposition at all, the common discount rate should not be too large.

You may wonder why I allow discounting at all. I wonder, too: no generation "should" be favored over any other. The usual scholarly excuse—which relies on the idea that there is a small fixed probability that civilization will end during any little interval of time—sounds farfetched. We can think of intergenerational discounting as a concession to human weakness or as a technical assumption of convenience (which it is). Luckily, very little of what I want to say depends on the rate of discount, which we can just imagine to be very small.

Given this discounting of future consumption, we have to imagine that our toy economy makes its investment and resource-depletion decisions so as to generate the largest possible sum of satisfactions over all future time. The limits to this optimization process are imposed by the pre-existing stock of resources, the initial stock of reproducible capital, the size of the labor force, and the technology of production.

This assumption of optimality is an embarrassing load to carry around. Its function is primarily to allow the semi-fiction that market prices accurately reflect scarcities. A similar assumption is implicit whenever we use ordinary GDP as a measure of economic well-being. In practice, no doubt, prices reflect all sorts of distortions arising from monopoly, taxation, poor information, and other market imperfections. In practice one can try to make adjustments to market prices to correct for the worst distortions. The conceptual points I want to make would survive. They are not to be taken literally in any case, but more as indicators of the sort of measurements we should be aiming at in principle.

Properly Charging the Economy for the Consumption of Its Resource Endowment. Now I come to the first major analytical step in my argument. If you look carefully at the solution to the problem of intergenerational resource allocation I have just sketched, you see that an excellent approximation of each single period's contribution to social welfare emerges quite naturally from the calculations. It is, in fact, a corrected version of net domestic product. The new feature is precisely a deduction for the net depletion of exhaustible resources. (I use the phrase "net depletion" because it is possible to extend this reasoning to allow for some discovery and development of new resources. In the pure case, where all discovery and development have already taken place, net and gross depletion coincide.)

The correct charge for depletion should value each unit of resource extracted at its net price, namely, its real value as input to production minus the *marginal* cost of extraction. As Hartwick has pointed out, if the marginal cost of mining exceeds average cost, which is what one would expect in an extractive industry, then the simple procedure of deducting the gross margin in mining (that is, the value of sales less the cost of extraction) will overstate the proper deduction and thus understate net product in the economy. If I may use the jargon of resource economics for a moment, the correct measure of depletion for social accounting prices is just the aggregate of Hotelling rents in the mining industry. That is the appropriate way to put a figure on what is taken from the ground in any given year, that year's withdrawal from the original endowment of nonrenewable resources.

This proposal presents two practical difficulties for national income accounting. The first is that observed market prices have to be corrected for the worst of the distortions I have just listed (that is, the distortion that would result from deducting the gross margin in mining—overstatement of the proper deduction and understatement of the net product in the economy). Making adjustments to market prices to correct for distortions is attempted routinely by the World Bank and other agencies in making project evaluations in developing countries. We seem to ignore the problem of

> It seems to me that the proper measurement of resource rents is exactly where the fund of knowledge embodied in an organization like RFF can find its application. This measurement would tell us something important about the true net product of our own economy. It should also be an input into policy decisions with a view to sustainability.

such distortions when we use our own national income accounts to study and judge the economies of advanced countries. If we are justified in that practice, the same casual treatment may be satisfactory in this context. (Not always, however: the large observed fluctuations in the price of oil cannot be accepted as indicating "true" values.) Either way, this is a surmountable problem.

I am not sure whether it is safe to be so casual about the second practical difficulty that my proposal for deducting net depletion of exhaustible resources presents for national income accounting. In principle, the proper measurement of resource rents requires the use of a numerical approximation to the marginal cost of mining. As I said, if marginal cost exceeds average cost by a lot, then taking the easy way out (just deducting the gross margin in mining) would entail a large error by overstating the depreciation of the resource stock. It seems to me that this is exactly where the fund of knowledge embodied in an organization like RFF can find its application. Tentative calculations for the main extractive industries would tell us something important about the true net product of our own economy. That would be important not merely because it would allow a more accurate evaluation of the path the economy has been following, but also, as you will see, because the measurement of resource rents should be an input into policy decisions with a view to sustainability.

Correcting National Accounts to Reflect Environmental Amenities. Pretty clearly, similar ideas should apply to a program of correcting the conventional national accounts to reflect environmental amenities. Much more attention has been lavished on environmental accounting than on resource accounting, and I have very little to add. Henry M. Peskin's work (much of which was done here at RFF) goes back to the early 1970s, and the Organisation for Economic Co-operation and Development, the World Bank, and the U.S. Department of Commerce are preparing a framework for integrating national income and environmental accounts. The sooner it happens the better. My only comment is a theoretical one. Without too much strain, it may be possible to treat environmental quality as a stock, a kind of capital that is "depreciated" by the addition of pollutants and "invested in" by abatement activities. In such cases the same general principles apply as to other forms of capital. The same intellectual framework will cover reproducible capital, renewable and nonrenewable resources, and environmental "capital."

The data problems may be altogether different, of course, especially when it comes to the measurement of benefits, a nicety that does not arise in the case of resource depletion. But the underlying treatment will follow the same rules. This counts for more than fastidiousness, I think. It would be a real achievement if it were to become a commonplace that capital assets, natural assets, and environmental assets were equally "real" and subject to the same scale of values, indeed the same bookkeeping conventions. Deeper ways of thinking might be affected.

> **The very logic of the economic theory of capital tells us how to construct a net national product concept that allows properly for the depletion of nonrenewable resources, and also for other forms of natural capital. Carrying out those instructions is far from easy... perhaps RFF could take the lead, as it has done with respect to environmental costs and benefits.**

That completes the first phase of my argument, so I will summarize briefly. The very logic of the economic theory of capital tells us how to construct a net national product concept that allows properly for the depletion of nonrenewable resources, and also for other forms of natural capital. Carrying out those instructions is far from easy, but that only makes the process more interesting. The importance of doing the work and doing it right is that theory underlines the basic similarity among all forms of capital, and that is a lesson worth learning. It will be reinforced by routine embodiment in the national accounts. Perhaps RFF could take the lead, as it has done with respect to environmental costs and benefits.

ANALYZING SUSTAINABLE PATHS FOR A MODERN INDUSTRIAL SOCIETY

Now I want to start down an apparently quite different path, but I promise that it will eventually link up with the unromantic measurement issues I have discussed so far, and will even reinforce the argument I have made.

I do not have to remind you that "sustainability" has become a hot topic in the last few years, beginning, I suppose, with the publication of the Brundtland Commission's report, *Our Common Future,* in 1987. As far as I can tell, however, discussion of sustainability has been mainly an occasion for the expression of emotions and attitudes. There has been very little analysis of sustainable paths for a modern industrial economy, so that we have little idea of what would be required in the way of policy and what sorts of outcomes could be expected. As things stand, if I express a commitment to sustainability, all that tells you is that I am unhappy with the modern consumerist life-style. If I pooh-pooh the whole thing, on the other hand, all you can deduce is that I am for business as usual. It is not a very satisfactory state of affairs.

Understanding What It Is That Must Be Conserved. If sustainability means anything more than a vague emotional commitment, it must require that something be conserved for the very long run. It is very important to understand what that something is: I think it has to be a generalized capacity to produce economic well-being.

It makes perfectly good sense to insist that certain unique and irreplaceable assets should be preserved for their own sake; nearly everyone would feel that way about Yosemite or, for that matter, about the Lincoln Memorial, I imagine. But that sort of situation cannot be universalized: it would be neither possible nor desirable to "leave the world as we found it" in every particular.

Most routine natural resources are desirable for what they do, not for what they are. It is their capacity to provide usable goods and services that we value. Once that principle is accepted, we are in the everyday world of substitutions and trade-offs.

For the rest of this talk, I will assume that a sustainable path for the national economy is one that allows every future generation the option of being as well off as its predecessors. The duty imposed by sustainability is to bequeath to posterity not any particular thing—with the sort of rare exception I have mentioned—but rather to endow them with whatever it takes to achieve a standard of living at least as good as our own and to look after their next generation similarly. We are not to consume humanity's capital, in the broadest sense. Sustainability is not always compatible with discounting the well-being of future generations if there is no continuing technological progress. But I will slide over this potential contradiction because discount rates should be small and, after all, there is technological progress.

> The duty imposed by sustainability is to bequeath to posterity not any particular thing—with rare exceptions such as Yosemite, for example—but rather to endow them with whatever it takes to achieve a standard of living at least as good as our own and to look after their next generation similarly.
> We are not to consume humanity's capital, in the broadest sense.

All that sounds bland, but it has some content. The standard of living achievable in the future depends on a bundle of endowments, in principle on everything that could limit the economy's capacity to produce economic well-being. That includes nonrenewable resources, of course, but it also includes the stock of plant and equipment, the inventory of technological knowledge, and even the general level of education and supply of skills. A sustainable path for the economy is thus not necessarily one that conserves every single thing or any single thing. It is one that replaces whatever it takes from its inherited natural and produced endowment, its material and intellectual endowment. What matters is not the particular form that the replacement takes, but only its capacity to produce the things that posterity will enjoy. Those depletion and investment decisions are the proper focus.

Outlining Two Key Propositions. Now it is time to go back to the toy economy I described earlier and to bring some serious economic theory to bear. There are two closely related logical propositions that can be shown to hold for such an economy. The first tells us something about the properly defined net national product, calculated with the aid of the right prices. At each instant, net national product indicates the largest consumption level that can be allowed this year if future consumption is never to be allowed to decrease.

To put it a little more precisely: net national product measures the maximum current level of consumer satisfaction that can be sustained forever. It is, therefore, a measure of sustainable income given the state of the economy—capital, resources, and so on—at that very instant.

This is important enough and strange enough to be worth a little explanation. How can this year's NNP "know" about anything that will or can happen in the future? The theorist's answer goes something like this. The economy's net product in any year consists of public and private consumption and public and private investment. (I am ignoring foreign trade altogether. Think of the economy as representing the world.) The components of investment, including the depletion of natural resources, have to be valued. That is where the "rightness" of the prices comes in. If the economy or its participants are forward-looking and far-seeing, the prices of investment goods will reflect the market's evaluation of their future productivity, including the productivity of the future investments they will make possible. The right prices will make full allowance even for the distant future, and will even take account of how each future generation will look at its future.

This story makes it obvious that everyday market prices can make no claim to embody that kind of foreknowledge. Least of all could the prices of natural resource products, which are famous for their volatility, have this property; but one could entertain legitimate doubts about other prices, too. The hope has to be that a careful attempt to average out speculative movements and to correct for the other imperfections I listed earlier would yield adjusted prices that might serve as a rough approximation to the theoretically correct ones. We act as if that were true in other contexts. The important hedge is not to claim too much.

16

While it is closely related to the proposition that NNP measures the maximum current level of consumer satisfaction that can be sustained forever, the second theoretical proposition I need is considerably more intuitive, although it may sound a little mysterious, too. Properly defined and properly calculated, this year's net national product can always be regarded as this year's interest on society's total stock of capital. It is absolutely vital that "capital" be interpreted in the broadest sense to include everything, tangible and intangible, in which the economy can invest or disinvest, including knowledge. Of course this stock of capital must be evaluated at the right prices. And the interest rate that capitalizes the net national product will generally be the real discount rate implicit in the whole story. Investment and depletion decisions determine the real wealth of the economy, and each instant's NNP appears as the return to society on the wealth it has accumulated in all forms. There are some tricky questions about wage incomes, but they are off the main track and I shall leave them unanswered.

Maintaining the Broad Stock of Society's Capital Intact. Something interesting happens when these two propositions are put together. One of them tells us that NNP at any instant is a measure of the highest sustainable income achievable, given the total stock of capital available at that instant. The other proposition tells us that NNP at any instant can be represented as that same stock of capital multiplied by an unchanging discount rate. Suppose that one goal of economic policy is to make investment and depletion decisions this year in a way that does not erode sustainable income. Then those same decisions must not allow the aggregate capital stock to fall. To use a Victorian phrase, preserving sustainability amounts to maintaining society's capital intact.

Let me say that in a slightly different way, speaking more picturesquely of generations rather than of instants or years. Each generation inherits a capital stock in the very broad and inclusive sense that matters. In turn, each generation makes consumption, investment, and depletion decisions. It enjoys its own consumption and leaves a stock of capital for the next generation. Of course, generations do not make decisions; families, firms, and governments do. Still, if all those decisions eventuate in a very large amount of current consumption, clearly the next generation might be forced to start with a lower stock of capital than its parents did. We now know that this is equivalent to saying that the new sustainable level of income is lower than the old one. The high-consumption generation has not lived up to the ethic of sustainability.

In the opposite case, consider a generation that consumes very little and leaves behind it a larger stock of capital than it inherited. That generation will have increased the sustainable level of income, and done so at the expense of its own consumption. Obviously that is what most past generations in the United States have done. Equally obviously, they were helped by ongoing technological progress. I have left that factor out of account, because it makes things too easy. It could probably be accommodated in the theoretical picture by imagining that there is a stock of technological knowledge that is built up by scientific and engineering research and depreciates through obsolescence. We know so little about that process that the formalization seems almost misleading. But the fact is very important.

A concern for sustainability implies a bias toward investment. That does not mean investment *über alles;* it means just enough investment to maintain the broad stock of capital intact. It does not mean maintaining intact the stock of every single thing; trade-offs and substitutions are not only permissible, they are essential. Unfortunately I have to make the limp statement that the terms on which one form of capital should be traded off against another are given by those adjusted prices—"shadow prices" we call them—and they involve a certain amount of guesswork. The guesswork has to be done; it cannot be avoided by defining the problem away. It is better that the guesswork be based on careful research than that the decision be fudged.

CONNECTING UP THE ARGUMENTS

Knowing What and How Much Should Be Replaced. Now I can connect up the two halves of my argument. Every generation uses up some part of the earth's original endowment of nonrenewable resources. There is no alternative. Not now anyway. Maybe eventually our economy will be based entirely on renewables. (The theory I have been using can be applied then too, with routine modifications.) Even so, there will be a long meanwhile. What should each generation give back in exchange for depleted resources if it wishes to abide by the ethic of sustainability? We now have an answer in principle. It should add to the social capital in other forms, enough to maintain the aggregate social capital intact. In other words, it should replace the used-up resources with other assets of equal value, or equal shadow value. How much is that? The shadow value of resource depletion is exactly the aggregate of Hotelling rents. It is exactly the quantity that should be deducted from conventional net national product to give a truer NNP that takes account of the depletion of resources. A research project aimed at estimating that deduction would also be estimating the amount of investment in other forms that would just replace the productive capacity dissipated in resource depletion. This is sometimes known as Hartwick's rule: a society that invests aggregate resource rents in reproducible capital is preserving its capacity to sustain a constant level of consumption.

Once again, I should mention that the same approach can be applied to environmental assets—the most complete treatment is by Karl-Göran Mäler—and to renewable resources—as in the work of John Hartwick. The environmental case is more complex, because even a stylized model of environmental degradation and rehabilitation is more complex than a model of resource depletion. The principle is the same, but the execution is even more difficult. Remember that even the simplest case offers daunting measurement problems.

Translating Sustainability into Policy. It is possible that the clarity brought to the idea of sustainability by this approach could lift the policy debate to a more pragmatic, less emotional level. But I am inclined to think that a few numbers, even approximate numbers, would be much more effective in turning discussion toward concrete proposals and away from pronunciamentos.

Suppose that the Department of Commerce published routinely a reasonable approximation to the "true" value of each year's depletion of nonrenewable resources. We could then say to ourselves: we owe to the future a volume of investment that will compensate for this year's withdrawal from the inherited stock.

We know the rough magnitude of this requirement. The appropriate policy is to generate an economically equivalent amount of net investment, enough to maintain society's broadly defined stock of capital intact. Of course, there may be other reasons for adding to (or subtracting from) this level of investment. The point is only that a commitment to sustainability is translated into a commitment to a specifiable amount of productive investment.

By the way, the same sort of calculation should have a very high priority in primary producing countries, the ones that supply the advanced industrial world with mineral products. They should also be directing their—rather large—Hotelling rents into productive investment. They will presumably want to invest more than that, because sustainability is hardly an adequate goal in poor countries. In this perspective, the cardinal sin is not mining; it is consuming the rents from mining.

> It is possible that the clarity brought to the idea of sustainability by this approach could lift the policy debate to a more pragmatic, less emotional level. But I am inclined to think that a few numbers, even approximate numbers, would be much more effective in turning discussion toward concrete proposals and away from pronunciamentos.

It goes without saying that this concrete translation of sustainability into policy leaves a lot of questions unanswered. The split between private and public investment has to be made in essentially political ways, like the split between private and public saving. There are other reasons for public policy to encourage or discourage investment, because there are social goals other than sustainability. One could hope for more focused debate as trade-offs are made more explicit.

I want to remind you again that environmental preservation can be handled in much the same way. It is a more difficult context, however, for several reasons. Many, though not all, environmental assets have a claim to intrinsic value. That is the case of the Grand Canyon or Yosemite National Park, as noted earlier.

The claim that a feature of the environment is irreplaceable, that is, not open to substitution by something equivalent but different, can be contested in any particular case, but no doubt it is sometimes true. Then the calculus of trade-offs does not apply. Useful minerals are in a more utilitarian category, and that is why I dealt with them explicitly.

Yet another difficulty is the deeper uncertainty about environmental benefits and costs. Marketed commodities, like minerals or renewable natural resources, are much simpler. I have admitted, fairly and squarely, how much of my argument depends on getting the shadow prices approximately right. Ordinary transaction prices are clearly not the whole answer; but they are a place to start. With environmental assets, not even that benchmark is available. I do not need to convince this audience that the difficulty of doing better does not make zero a defensible approximation for the shadow price of environmental amenity. I think the correct conclusion is the one stated by Karl-Göran Mäler: that we are going to have to keep depending on physical and other special indicators in order to judge the economy's performance with regard to the use of environmental resources. Even so, the conceptual framework should be an aid to clear thinking in the environmental field as well.

Maybe this way of thinking about environmental matters offers a way out of a dilemma facing less developed countries. The dilemma arises because they sometimes find that the adoption of developed-country environmental standards makes local industries uncompetitive in world markets. The poor countries then seem to have a choice between cooperating in the degradation of their own environment or acquiescing in their own poverty. At least when pollution is localized, the resolution of the dilemma appears to be a controlled trade-off between an immediate loss of environmental amenity and a gain in future economic well-being. Temporary acceptance of less-than-the-best environmental conditions can be made more palatable if the "rents" from doing so are translated into productive investment. Higher incomes in the future could be spent in part on environmental repair, of course, but it is general well-being that counts ultimately.

Notice that I have limited this suggestion to the case of localized pollution. When poor countries in search of their own economic goals contribute to global environmental damage, much more difficult policy questions arise. Their solution is not so hard to see in principle, but the practical obstacles are enormous. In any case, I leave those problems aside.

CONCLUDING COMMENTS

That brings me to the end of my story. I have suggested that an innovation in social accounting practice could contribute to more rational debate and possibly to more rational action in the economics of nonrenewable resources and the approach to a sustainable economy. There is a trick involved here, and I guess I should confess what it is. In a complex world, populated by people with diverse interests and tastes, and enmeshed in uncertainty about the future (not to mention the past), there is a lot to be gained by transforming questions of yes-or-no into questions of more-or-less. Yes-or-no lends itself to stalemate and confrontation; more-or-less lends itself to trade-offs. The trick is to understand more of what and less of what. This lecture was intended to make a step in that direction.